Understanding Mathematics From First Principles

Why Numbers Work, How Proofs Matter, and the Foundations Every Learner Should Know

Honest Earning

An independent mathematics educator dedicated to making complex ideas accessible to every learner

Table of Contents

"Honest Earning" is a pen name. The author writes under this name to keep the focus on the mathematical content rather than personal identity.

This is an entirely original work. All explanations, examples, and proofs are the independent creation of the author.

Published by Honest Earning Publishing, India. First Edition, 2026.

This book is for educational purposes. Suitable for readers aged 13 and above.

ISBN (Paperback): 978-93-6068-554-6

For every student who ever whispered "I'm just not a math person" under their breath in a classroom.

You were wrong. This book is the proof.

About the Author

Honest Earning is the pen name of an independent mathematics educator dedicated to making complex mathematical ideas accessible to every learner. Years of working with students — teenagers stuck on algebra, adults rebuilding their confidence after decades away from a classroom — led to one conviction: almost nobody actually dislikes math. What they dislike is being confused by it, and then being told that's their own fault.

That confusion is almost always traceable. A rule got handed down without a reason. A shortcut got memorized before it made sense. A "why" got waved away as too obvious to bother explaining.

This book tries to undo that. It moves slowly, skips nothing, and asks you to think rather than memorize. The goal isn't to race through a syllabus — it's to give you the actual experience of understanding something in mathematics, all the way down. Once you've felt that, it's hard to settle for anything less.

Preface

This book started as an answer to a single question a student once asked me after class: "Why does a negative times a negative equal a positive? I know it does. I just don't know *why*." I gave the usual analogy — the enemy of my enemy is my friend — and watched it land the way analogies do: politely accepted, not actually believed.

That question stayed with me. Over the following months I went looking for a real answer, not a story, and found one — the proof in Chapter 7 of this book. It took about a page of algebra, built entirely from properties so simple they're barely ever mentioned. What struck me wasn't the proof itself; it was how *available* it had been the whole time, sitting just underneath a rule millions of students memorize every year without ever being shown it.

That became the project: take the handful of ideas that make arithmetic actually work, and lay them out properly, in order, with nothing skipped and nothing hand-waved. Not a faster way to pass a test. A genuine, patient answer to "why."

If you've ever had a "why" question about math that never got answered — this book was written for that question, whatever it turns out to be.

— Honest Earning

Introduction: Why You Are Not Bad at Mathematics

> *"Mathematics is not about numbers, equations, computations, or algorithms: it is about understanding." — William Paul Thurston*

There's a moment almost everyone remembers. Somewhere between ten and sixteen years old, math stopped making sense. One year you were fine, the next you were lost, and somewhere in there you quietly decided you just "weren't a math person." That story is wrong, and this book exists to prove it.

Here's the pattern I've seen over and over, working with people who were sure they were hopeless at math: it was almost never about intelligence. It was a gap — one idea, taught once, that never quite landed — and every topic built on top of that gap felt shakier and shakier, like a house built on a step that isn't level. Close the gap, and the wobble disappears. Suddenly the thing that felt like a foreign language starts reading like plain logic.

> REAL EXAMPLE
>
> Think about learning to drive. At first you're just following steps: mirror, signal, turn the wheel this much. But at some point it clicks — you *understand* how the car responds — and after that you're not following steps anymore, you're just driving. Math foundations work the same way. Once the underlying logic clicks, you stop following recipes and start reasoning.

What This Book Promises

Read carefully, work through the questions at the end of each chapter, and you will come away with real mathematical understanding — not just facts you can recite, but reasons you can explain. You'll feel the difference between "I memorized that negative times negative is positive" and "I can show you exactly why it has to be that way."

The scope here is narrow on purpose: whole numbers, negative numbers, and the logic that connects them. Going deep on a small, solid foundation is worth far more than skimming the surface of a hundred topics.

Who This Is For

This book is for anyone who wants to actually *get* math, not just get through it.

Maybe you're an adult who studied this stuff in school, forgot most of it, and have had "why is negative times negative positive?" sitting unanswered in the back of your mind for twenty years. Maybe you're a student who can solve the equations but has no idea why the steps work. Maybe you're a teacher looking for a fresh way to explain the basics. Whoever you are, if you're willing to slow down and think, this book is built for you.

How This Book Is Organized

The first two chapters set up why foundations matter and where numbers actually came from.

Chapters 3 and 4 pin down exactly what natural numbers and integers are. Chapter 5 lays out the rules of arithmetic — and why they're rules, not accidents. Chapters 6 and 7 build up the tools of proof, ending with a full, rigorous explanation of why a negative times a negative is positive. Chapters 8 and 9 put those tools to work. Chapter 10 ties it together with a satisfying final result about even and odd numbers.

Every chapter ends with three things: **Key Takeaways** (the short version), **Reflection Questions** (to make you think, not just read), and **Action Steps** (things to actually do — get a pencil out).

> PLAIN ENGLISH
>
> Throughout the book, boxes like this one translate technical language into everyday words. If a term feels unfamiliar, check the nearest one of these.

> **KEY INSIGHT**
> Boxes like this one flag the single most important idea on the page — the one worth pausing on before you move forward.

Get a notebook before you start. Math isn't a spectator sport — when a problem shows up, stop and try it before reading the answer.

A Word on Getting Stuck

You will get stuck at some point in this book. That's not a sign you're bad at math — it's a sign you've bumped into a real idea, one that takes effort to absorb. When that happens, don't skip ahead. Slow down. Reread the sentence you're least sure about. Try the example yourself before checking the answer. That slow, effortful process *is* how understanding gets built — it can't be rushed, and it isn't reserved for "math people." It's a skill, and you're about to practice it.

Chapter 1: The Power of Foundations

> *"To understand something is to be delivered of it." — Baruch Spinoza*

The Skyscraper and the Soil

Civil engineers know something that math students often miss: how tall a building can go depends entirely on what's underneath it. A skyscraper on loose soil can't rise. The same tower on solid bedrock can reach hundreds of floors. Mathematical foundations work exactly the same way. The arithmetic rules in Chapter 5 look almost too simple to bother stating — but they're the bedrock underneath algebra, calculus, statistics, and everything else. Shaky foundations mean everything built on top wobbles too.

Most students experience math as a pile of disconnected tricks: here's how you add fractions, here's how you solve for x, here's how you find a derivative. Each one gets taught in isolation, drilled until it's automatic, then dropped for the next trick. That produces people who can follow a recipe but can't cook — who can execute a method but can't tell you why it works, or what to do when the problem doesn't quite match the recipe. This book does the opposite. We spend real time on foundations, not because they're hard, but because understanding them changes everything that comes after.

Knowing vs. Understanding

There's a real difference between knowing a fact and understanding it. Knowing means you can recite it. Understanding means you know *why* it's true, where it comes from, and what would have to be different for it to be false.

> REAL EXAMPLE
>
> Say you know the formula for the area of a triangle: half times base times height. That's knowing. Understanding is being able to draw a rectangle around the triangle, see that the

> triangle is exactly half of it, and derive the formula yourself — so that even if you forgot the formula tomorrow, you could rebuild it in thirty seconds.

> **KEY INSIGHT**
> Understanding is durable in a way that knowing isn't. You can forget a formula overnight. You basically can't "forget" an understanding — it leaves a trace that makes relearning fast and applying it to new problems possible.

The Rote-Learning Trap

The default way math gets taught is: here's the rule, here's how to apply it, practice until it's automatic. That's not entirely wrong — being fluent in basic procedures is genuinely useful. But it becomes a trap when it replaces understanding. Students who only learned by rote find their skills evaporate the moment practice stops, get thrown by any problem that looks slightly different from the ones they drilled, and build advanced topics on a foundation that was never actually there.

The fix isn't more practice. It's going back and finding the actual gap — usually somewhere in the properties of arithmetic, the definitions of the number systems, or what a mathematical proof even is. That's exactly where this book starts.

What to Expect

This book covers a narrow slice of math — whole numbers and integers — but covers it properly. By the end, you'll understand *why* the rules of arithmetic are what they are, you'll have written real mathematical proofs, and you'll have solved some genuinely elegant puzzles. The Roadmap at the back gives you a clear path for what to study next.

Key Takeaways

- Mathematical foundations aren't warm-up material — they're what everything else stands on.
- Knowing and understanding are different in kind, not just in degree. Understanding is durable; knowing is brittle.

- Rote learning builds fluency, not understanding. Understanding means knowing *why* the rules hold.

Reflection Questions

1. Name one math rule you know but couldn't explain if someone asked "why." What would it take to actually understand it?
2. Think of something you learned by rote and later actually understood. What changed?
3. Why might deeply understanding ten ideas beat superficially knowing a hundred?

Action Steps

4. Write three sentences about your current relationship with math. Date it. Reread it when you finish this book.
5. Using only the idea that multiplication is repeated addition, try to prove from scratch that $(a+b)^2 = a^2 + 2ab + b^2$.
6. Get a notebook dedicated to this book. Don't read a chapter without it nearby.

Chapter 2: The Story of Counting — How Numbers Were Born

> *"The art of doing mathematics consists in finding that special case which contains all the germs of generality."* — *David Hilbert*

A World Without Numbers

Try to imagine a day with no numbers at all — no clock, no prices, no phone number, no score. It's hard, because modern life is soaked in numbers. But for almost all of human history, numbers didn't exist. Modern humans have been around for roughly 300,000 years; writing is only about 5,000 years old. For the overwhelming majority of our history as a species, nobody was counting anything.

That doesn't mean those people were unintelligent — they built shelters, made fire, hunted in coordinated groups, made sophisticated art. They just didn't have numbers. So what problem did counting actually solve?

The Problem Numbers Solve

Counting solves one specific problem: comparing things reliably across time. Can you eyeball whether your grocery bag has as many apples in it now as it did an hour ago, without counting? Not really — unless both moments are right in front of you. But if you counted seven apples earlier, and now you count five, you know instantly and exactly what changed.

> REAL EXAMPLE
>
> This is why every business runs on counting: how many days of inventory are left, how much cash is in the register, how many seats are still unsold on a flight. Counting is a tool for bridging what you can see right now and information about a moment you can no longer see.

The Earliest Evidence

The oldest physical evidence of counting comes from two remarkable objects. The Lebombo Bone, found in a cave in Swaziland, is around 43,000 years old and carries 29 notches that may track a lunar cycle. The Ishango Bone, from the Democratic Republic of Congo and roughly 20,000 years old, shows notch patterns some researchers read as early multiplication and prime numbers. Whatever the full story, by 20,000 years ago people were deliberately recording structured numerical patterns.

Tally Marks and Their Limits

The simplest number system is the tally: one mark per item. Five sheep, five marks. It's dead simple and needs zero abstraction — but it breaks down fast. To represent fifty things you need fifty marks, which is as unwieldy as the pile itself. A row of twenty-two marks is easy to miscount as twenty-three. And you can't easily multiply or divide a pile of tally marks.

> PLAIN ENGLISH
> "Arithmetic" just means the four operations: adding, subtracting, multiplying, dividing. A

> number system that "supports arithmetic" lets you do these operations on the symbols themselves, without going back to the physical pile of objects.

The Leap to Abstraction

The single biggest turning point in the history of math wasn't a formula — it was realizing that numbers are things in their own right, separate from whatever's being counted. Before this leap: three sheep, three stones, and three days were three unrelated things. After: they all share a property called "three-ness," and the number 3 can be studied on its own, without any sheep, stones, or days attached to it.

> **KEY INSIGHT**
> The jump from "three sheep" to "the number 3" is arguably the single most important idea in the history of math. Every other idea in this book is built on top of it.

Positional Notation and the Revolution of Zero

The breakthrough that gives us modern numbers is *positional notation*: a symbol's value depends on where it sits. In "345," the 3 means three hundred, the 4 means forty, and the 5 means five — not because of anything inherent in those digits, but purely because of position. The same digit, 3, means three, thirty, or three hundred depending on where you put it. This system — the Hindu-Arabic numerals we still use — was developed in India; the mathematician Brahmagupta gave the first systematic treatment of arithmetic with zero in the 7th century CE. It later spread through the Islamic world into Europe, gradually replacing Roman numerals over several centuries.

> **PLAIN ENGLISH**
> "Positional" means position determines value. "Decimal" means base ten — ten symbols, 0 through 9, probably because humans have ten fingers. Zero isn't just "nothing" here; it's a full number with its own arithmetic behavior, and

> that's what makes the whole positional system work.

Why Zero Changed Everything

Zero is the strangest, most important piece of this system. Without it as a placeholder, you can't tell 3 from 30 — you need a symbol that means "nothing goes here." But zero is also a genuine number: add it to anything and nothing changes; multiply anything by it and you get zero. Those two properties are formalized in Chapter 5.

Key Takeaways

- Numbers were invented to solve a real problem: comparing quantities reliably across time.
- Tally marks were the first number system; their limits pushed people toward better ones.
- The critical leap was abstraction — numbers as things in their own right, not tied to specific objects.
- The Hindu-Arabic positional decimal system, developed in India, is one of history's great intellectual inventions.

- Zero is a number with real arithmetic properties, not just an absence.

Reflection Questions

7. What does it mean for a number to be an "abstract object," and how is that different from a physical object like a stone?
8. Try multiplying 47 by 23 using Roman numerals. What does the struggle tell you about the link between notation and arithmetic?
9. Brahmagupta gave the first systematic rules for zero, but got division by zero wrong. What does that tell you about how math actually develops?

Action Steps

10. Look up Brahmagupta's rules for zero from his *Brahmasphutasiddhanta*. Which one is wrong, and why, in your own words?
11. Find an image of the Ishango Bone. Do you find the "evidence of prime numbers" argument convincing? Write a short paragraph on your view.
12. Explain to a twelve-year-old, in one page, why the decimal system beats Roman numerals for actual arithmetic.

Chapter 3: Natural Numbers and the Infinity Argument

> *"The infinite! No other question has ever moved so profoundly the spirit of man." — David Hilbert*

Defining the Natural Numbers

Math is precise about its definitions in a way few other subjects are — a term means exactly what its definition says, no more, no less. That precision is what makes mathematical reasoning trustworthy. So let's start with one.

```
N = { 1, 2, 3, 4, 5, … }
```

The curly braces mean this is a *set* — a collection of distinct things with a clear rule for who's in and who's out. The "..." means the pattern keeps going forever. N has no last element; it's infinite.

> PLAIN ENGLISH
>
> A set is just a collection with a clear membership rule. Three dots always mean "the obvious pattern goes on forever." N = {1, 2, 3, ...} is every positive whole number, with no end.

Note: some books include 0 in the natural numbers. This one follows the common number-theory convention: natural numbers start at 1. Zero shows up in the integers, in Chapter 4.

We write **a ∈ S** to mean "a is in S," and **a ∉ S** to mean "a is not in S." So 5 ∈ N (5 is natural), but 0 ∉ N and −3 ∉ N under our convention.

Is There a Largest Natural Number?

Here's a question that sounds almost too simple: is there a biggest natural number? Most people's gut says no — but "you can always add one more" is a feeling, not a proof. Let's actually prove it, using a technique called **proof by contradiction**, which will be the workhorse method of this whole book.

> PLAIN ENGLISH
> Proof by contradiction: assume the *opposite* of what you're trying to prove. Follow where that leads. If it leads somewhere clearly impossible,

> then the assumption must have been wrong — which means your original claim is true.

Theorem: N has no largest element.

Proof: Suppose it did, and call it M. By assumption, M is bigger than or equal to every natural number. Now look at M + 1. Since M and 1 are both natural numbers, so is their sum — but M + 1 is bigger than M. We've just found a natural number larger than the "largest" one, which is a contradiction. So the assumption was false: N has no largest element. ∎

> REAL EXAMPLE
>
> A savings account with compounding interest never hits a ceiling on its own — you can always imagine one more deposit, one more year, one more dollar. There's no "final" balance where growth is forced to stop. That's the same idea as "no largest natural number," just wearing a bank statement instead of a proof.

What the Proof Does *Not* Say

It says: for any specific number you name, there's a bigger one — N is unbounded. It does *not* say N "reaches" some endpoint called infinity. Infinity isn't a natural number; it's the idea of having no bound at all. A lot of math mistakes come from treating infinity like an enormous number instead of what it actually is.

Some Infinities Are Bigger Than Others

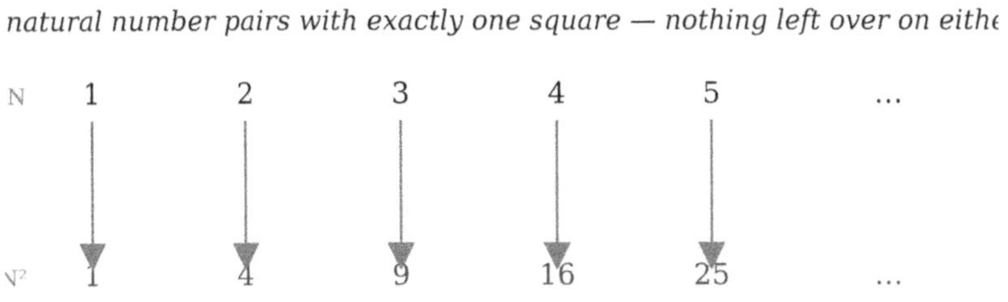

Cantor's pairing: every natural number matches exactly one square, with none left over on either side.

The ancient Greeks were deeply uneasy with the idea of a *completed* infinity — a whole, finished infinite collection, as opposed to just "you can always keep going." That discomfort lasted for over two thousand years, until the German mathematician Georg Cantor tackled it head-on in the late 1800s.

Cantor noticed something bizarre: you can pair up every natural number with its square (1↔1, 2↔4, 3↔9, ...) with nothing left over on either side — even though the squares are clearly a smaller *part* of the naturals. He resolved the apparent paradox by redefining what "same size" means for infinite sets: two sets are the same size if you can pair up their elements one-to-one, full stop. By that definition, the naturals and their squares genuinely are the same size. "The whole is bigger than the part" — true for finite collections, false for infinite ones.

Cantor then showed something even stranger: the real numbers (all decimals) *can't* be paired up one-to-one with the naturals. There are, in a precise sense, more real numbers than natural numbers — infinities come in different sizes. The idea was so radical that some of Cantor's contemporaries attacked him personally for it; he died in 1918 having been vindicated by almost nobody in his own lifetime. Today his work is considered foundational.

Mathematical Induction: The Other Face of Infinity

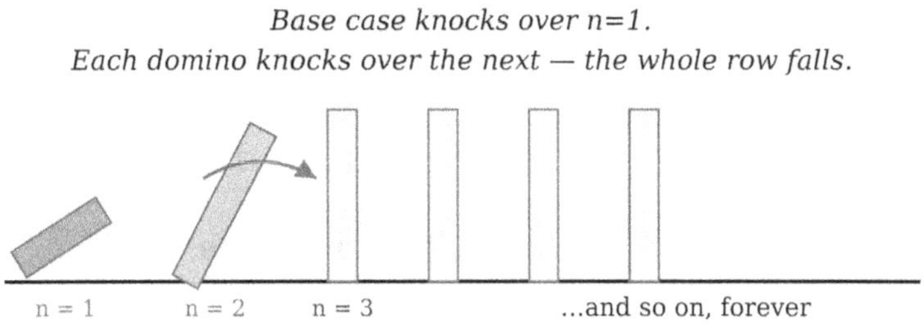

Induction, pictured: the base case tips the first domino, and each domino guarantees the next one falls.

There's a second major tool for reasoning about all natural numbers at once: **induction**. The idea: show a property holds for 1, then show that whenever it holds for some number n, it also holds for n+1. Conclude it holds for every natural number.

> REAL EXAMPLE
>
> Think of a long row of dominoes. If you know the first one falls, and you know each domino knocks

> over the next one, you know the whole row falls — without ever pushing every single domino yourself. That's induction.

Example: the sum 1 + 2 + 3 + ... + n always equals n(n+1)/2.

Base case (n=1): 1 = 1(2)/2 = 1. ✓

Inductive step: assume it holds for n. Then 1+2+...+n+(n+1) = n(n+1)/2 + (n+1) = (n+1)(n+2)/2 — exactly the formula for n+1. ✓

By induction, the formula holds for every natural number. (This is the same result we'll meet again in Chapter 9, credited to a young Carl Friedrich Gauss.)

Key Takeaways

- N = {1, 2, 3, ...} is the infinite set of natural numbers, precisely defined.
- a ∈ S means "a belongs to S"; a ∉ S means it doesn't.

- Proof by contradiction: assume the opposite, find an impossibility, conclude the original claim is true.
- N has no largest element — proven, not just assumed.
- Infinity isn't a number, and not all infinities are the same size.

Reflection Questions

13. Without looking back, try to reconstruct the "no largest natural number" proof from memory. Where do you hesitate, and why?
14. Is N closed under subtraction? Multiplication? Division? Justify each answer.
15. Cantor showed some infinities are bigger than others. Does that match or clash with your gut sense of "infinite"?

Action Steps

16. Write out the "no largest natural number" proof in your own words, one justified step per line.
17. Prove by contradiction that there's no smallest positive fraction. (Hint: if p/q were smallest, what about p/(2q)?)
18. Spot three moments in ordinary life where you used contradiction-style reasoning without naming it. Write each as: assumption → consequence → contradiction → conclusion.

Chapter 4: When Subtraction Broke Everything — The Birth of Integers

> *"The introduction of negative numbers was one of the most difficult steps in the history of mathematics." — Morris Kline*

The Accounting Problem

Natural numbers are great for counting things you actually have. They're useless for things you *owe*. Say you have five bags of rice and a buyer takes eight — how many do you have now? The honest answer is "negative three," but natural numbers have no way to say that.

> REAL EXAMPLE
>
> This shows up constantly: an overdrawn bank account, a temperature below zero, a floor below ground level in an elevator, a business running a loss for the quarter. All of these are natural, everyday situations that natural numbers simply cannot describe.

The Historical Struggle

Mathematicians didn't accept negative numbers overnight. Chinese and Indian mathematicians were using them for debt calculations as early as the 7th century CE, but European mathematicians resisted for centuries longer. René Descartes called negative solutions to equations "false" roots. Blaise Pascal wrote that nothing was more absurd than subtracting a bigger number from a smaller one.

These weren't unintelligent people — the resistance was a genuine philosophical hurdle. You can't hold negative three sheep in your hands. Accepting that a number could represent a debt, or a direction, rather than a physical pile of things, was a real conceptual leap, the same kind we saw with abstraction in Chapter 2. Once mathematicians made that leap, the payoff was enormous: negative numbers unlock algebra and describe huge swaths of the physical world.

Defining the Integers

The fix is to extend N by adding zero and the negative whole numbers. The result is the integers:

```
Z = { …, -4, -3, -2, -1, 0, 1, 2, 3, 4,
… }
```

(Z comes from the German *Zahlen*, "numbers.") The integers stretch in both directions forever. Every natural number is an integer, but not every integer is a natural number — we write **N ⊂ Z** ("N is contained in Z").

PLAIN ENGLISH

⊂ means "is contained in." N ⊂ Z means everything in N is also in Z, but Z has extra members too — zero and the negatives. Think of N as a smaller club inside the bigger club Z.

A Critical Warning: Division by Zero

IMPORTANT

Division by zero is undefined. Not zero, not infinity — undefined, full stop. Here's why: a ÷ b is defined as "the number c such that b × c = a." So a ÷ 0 = c would mean 0 × c = a. But 0 × c is always 0, no matter what c is. So if a isn't 0, no c

> can work — there's no answer. And if a *is* 0, every c works — there's no *unique* answer either way. Division by zero simply has nowhere to land. Whenever this book writes p/q, assume $q \neq 0$.

Why This Extension Matters

Extending N to Z makes subtraction complete: for any two integers a and b, $a - b$ is always an integer. Before, you could only subtract sometimes; now you always can. The equation $x + 5 = 3$ has no solution in N, but it has the solution $x = -2$ in Z. This is the pattern behind every number-system upgrade in history: a problem shows up that the current system can't solve, so a bigger system gets defined that solves it while keeping everything the old system could already do.

Ordering and the Number Line

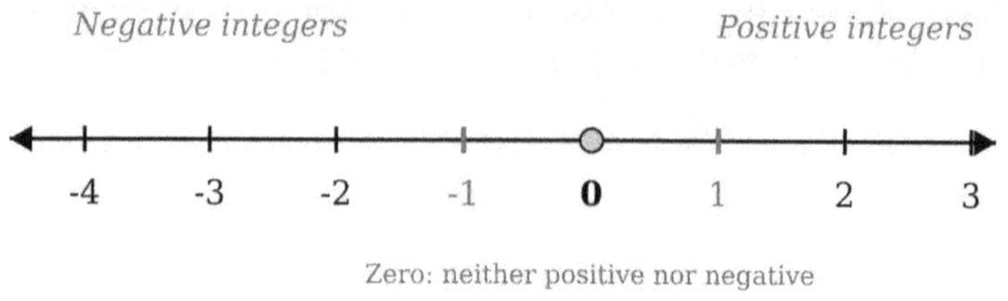

The integer number line: negatives to the left of zero, positives to the right.

The integers line up naturally on a number line: negatives to the left, zero in the middle, positives to the right. For any two integers a and b, exactly one of these is true: $a < b$, $a = b$, or $a > b$ — the *trichotomy law*.

One rule to remember: if $a < b$ and you multiply both sides by a positive number, the inequality stays the same direction. But multiply both sides by a *negative* number, and it flips: $a < b$ becomes $ac > bc$. That flip is a direct consequence of the sign rules we'll prove in Chapter 7, and it's one of the most common sources of algebra mistakes — precisely because most students memorize the flip instead of understanding where it comes from.

Key Takeaways

- Z = {..., −2, −1, 0, 1, 2, ...} extends N with zero and the negative integers.
- N ⊂ Z: every natural number is an integer, but not vice versa.
- The integers make subtraction complete.
- Division by zero is undefined — not zero, not infinity.
- Negative numbers were historically resisted even by mathematicians who understood them well.

Reflection Questions

19. Walk through the division-by-zero argument yourself: if a ÷ 0 = c, then 0 × c = a. What breaks in each case?
20. Name three real-world quantities that are naturally negative. What does the negative sign mean physically in each case?
21. Descartes called negative roots "false." What would it even mean for a number to be "true" or "false"?

Action Steps

22. Draw a number line from −10 to 10. Mark at least five real quantities on it — at least one

negative, zero, and two positive — using genuinely different kinds of things.

23. Look up Brahmagupta's original (incorrect) rule for dividing by zero, and explain precisely why it fails using this chapter's argument.
24. Write a short, formal proof that $N \subset Z$.

Chapter 5: The Properties of Arithmetic — Rules That Are Not Arbitrary

> *"In mathematics the art of asking questions is more valuable than solving problems."* — *Georg Cantor*

Why Bother Stating the Obvious?

Fair question: why write down that 3 + 5 = 5 + 3? Everyone already knows that. Two reasons. First, "obviously true" isn't the same as "proved" — math needs to be explicit about what it's assuming, so we know exactly what our reasoning rests on. Second, these properties *aren't* universal. Matrix multiplication doesn't commute ($AB \neq BA$ in general). Knowing precisely what each property says is what lets you spot when it applies and when it doesn't.

PLAIN ENGLISH

An *axiom* is a statement we accept without proof, as a starting point. A *theorem* is something we prove from the axioms. The nine properties below are the axioms for integer arithmetic in this book.

Property 1: Commutativity of Addition

For all a, b in Z: a + b = b + a

Order doesn't matter for addition. It does matter for subtraction ($3-5 \neq 5-3$) and division ($8 \div 4 \neq 4 \div 8$).

REAL EXAMPLE

Loading a grocery cart with a apples then b oranges gets you the same cart as b oranges then a apples. The total doesn't care about the order you added things in.

Property 2: Additive Identity

For all a in Z: a + 0 = 0 + a = a

Adding zero changes nothing. Zero is addition's "do nothing" number.

Property 3: Additive Inverse

For all a in Z, there is -a such that:
a + (-a) = 0

Every integer has an opposite that cancels it out to zero. Subtraction is *defined* using this: a − b means a + (−b).

> REAL EXAMPLE
> You spend ₹500, then earn ₹500 back. Net change: zero. The earning is the additive inverse of the spending.

Property 4: Associativity of Addition

```
For all a, b, c in Z:  (a + b) + c = a + (b + c)
```

How you *group* a chain of additions doesn't change the answer. (Subtraction isn't associative: (8−3)−2 = 3, but 8−(3−2) = 7 — try it.)

Property 5: Commutativity of Multiplication

```
For all a, b in Z:  a × b = b × a
```

> REAL EXAMPLE
> A rectangular room tiled a tiles wide and b tiles long uses the same number of tiles as one that's b wide and a long — just turned sideways. Same tiles, same count.

Property 6: Multiplicative Identity

```
For all a in Z:  a × 1 = 1 × a = a
```

Multiplying by 1 changes nothing.

Property 7: Multiplicative Annihilator

```
For all a in Z:  a × 0 = 0 × a = 0
```

We can actually derive this rather than just asserting it: $a \times 0 = a \times (0+0) = a \times 0 + a \times 0$ (by distributivity, Property 9 below). Call this value x. Then $x = x + x$, so subtracting x from both sides gives $0 = x$. So $a \times 0 = 0$.

Property 8: Associativity of Multiplication

```
For all a, b, c in Z:  (a × b) × c = a ×
(b × c)
```

A box with sides a, b, and c has the same volume no matter which two sides you multiply first — the box itself hasn't changed.

Property 9: Distributivity

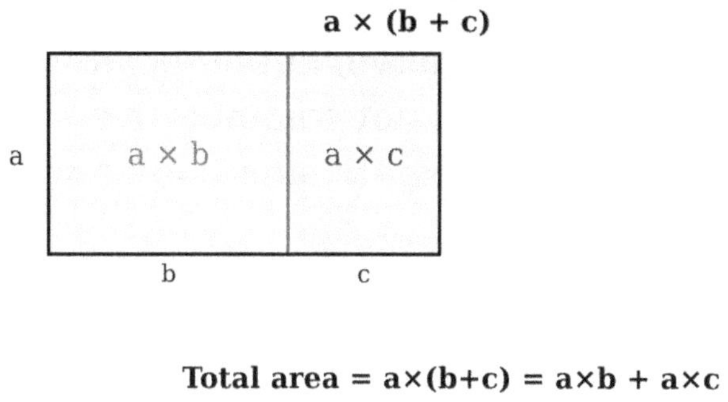

A rectangle split into two parts: the picture behind a×(b+c) = a×b + a×c.

For all a, b, c in Z: a × (b + c) = a × b + a × c

This is the most important of the nine — it's the bridge connecting addition and multiplication, and without it, algebra as we know it doesn't work.

> REAL EXAMPLE
>
> Splitting a restaurant bill three ways where everyone orders the same ₹200 starter plus their own main: 3 × (200 + main) = 3×200 + 3×main. Distributing the multiplication over the addition gives the same total either way you compute it.

Numerical check: 3×(4+5) = 3×9 = 27, and (3×4)+(3×5) = 12+15 = 27. ✓

Where These Properties Fail

- Subtraction: not commutative, not associative.
- Division: not commutative, not associative.
- Matrix multiplication: not commutative.

This is exactly *why* mathematicians bother writing these properties down explicitly — they're features of integer arithmetic, not universal truths, and each new mathematical setting has to be checked on its own terms.

Key Takeaways

- Nine properties form the foundation of integer arithmetic.
- Commutativity holds for addition and multiplication — not subtraction or division.
- Zero is addition's identity and multiplication's "annihilator"; one is multiplication's identity.
- Distributivity connects addition and multiplication and underlies all of algebra.
- These properties can fail in other mathematical settings — always check, never assume.

Reflection Questions

25. Derive $a \times 0 = 0$ yourself from distributivity and the additive identity, writing out every step.

26. Does addition distribute over multiplication? Test whether a + (b×c) equals (a+b)×(a+c). What do you find?
27. Find two 2×2 matrices A and B where AB ≠ BA. What does that tell you about how "obvious" rules can be context-specific?

Action Steps

28. Draw the rectangle-splitting picture for distributivity on paper, label every piece, and explain it out loud as if teaching someone who's never seen it.
29. Test associativity for subtraction using 10, 4, and 2. Compute (10−4)−2 and 10−(4−2), confirm they differ, and write a sentence explaining what that means.
30. Look up the definitions of a mathematical "group" and a "ring." Which of the nine properties does each require?

Chapter 6: Proof by Contradiction — The Art of Logical Argument

> *"A proof is a proof. What kind of proof? It's a proof. A proof is a proof, and when you have a good proof, it's because it's proven." — Jean Chrétien*

Revisiting Contradiction

We first used proof by contradiction in Chapter 3, to show N has no largest element. Now let's look closely at the technique itself, because its next job is one of the most famous proofs in all of mathematics.

> PLAIN ENGLISH
>
> *Reductio ad absurdum* is the Latin name for proof by contradiction: "reduction to the absurd." Take the opposite of your claim, follow it to its logical end, and show that end is absurd. Since the logic was sound, the starting assumption must have been the problem.

An Everyday Analogy

A colleague swears they emailed you an important file on Tuesday. You check your inbox — nothing between Monday and Thursday. You reason: if it had been sent Tuesday, it would be sitting in my inbox. It isn't. So it wasn't sent Tuesday.

REAL EXAMPLE

That's proof by contradiction, in the wild. Claim: "the email was sent." Assume the opposite: "it wasn't received, but it was sent." Consequence: it should be in the inbox. Contradiction: it isn't there. Conclusion: the claim (as stated) is false. Mathematical proofs are the same structure — the only difference is every step has to be airtight, justified by a definition, an axiom, or something already proven.

The Irrationality of √2: Setup

We're about to prove one of the most famous results in math history: √2 cannot be written as a fraction. Legend has it this discovery deeply disturbed the ancient Pythagoreans, who believed every quantity could be expressed as a ratio of whole numbers.

PLAIN ENGLISH

A *rational* number is anything expressible as a fraction p/q, with p and q whole numbers and q not zero — like 3/4 or 6 (which is 6/1). An

> *irrational* number can't be written that way. The claim is that $\sqrt{2}$ is irrational.

The Irrationality of $\sqrt{2}$: Proof

Theorem: $\sqrt{2}$ is irrational — there's no fraction p/q equal to it.

Proof. Assume the opposite: $\sqrt{2} = p/q$, written in lowest terms (so p and q share no common factor). Squaring both sides:

$$2 = p^2/q^2 \quad \Rightarrow \quad p^2 = 2q^2$$

So p^2 is even. A quick lemma: if p^2 is even, p itself must be even (if p were odd, p^2 would come out odd too — try it with $p = 2k+1$). So p is even; write $p = 2m$. Substituting:

$$(2m)^2 = 2q^2 \quad \Rightarrow \quad 4m^2 = 2q^2 \quad \Rightarrow \quad q^2 = 2m^2$$

So q^2 is even too, which by the same reasoning means q is even. But now both p and q are even — they share a factor of 2, which contradicts our "lowest terms" assumption. So the original assumption was false: √2 is irrational. ∎

> **KEY INSIGHT**
> Every step here is earned. The "if p^2 is even, p is even" fact was proved separately as a supporting lemma. Breaking a big proof into smaller, individually-justified pieces is standard practice, in math and in any careful argument.

Circular Reasoning: The Fatal Error

The most common way a proof goes wrong is circular reasoning — quietly using the thing you're trying to prove as part of the argument for it. A circular "proof" says, in effect, "P is true because Q is true, and Q is true because P is true," which proves nothing at all.

Whenever you read or write a proof, ask at each step: *what is this step assuming, and have I actually established that already?* In the √2 proof, nothing used depends on √2 already being irrational — it's clean.

Key Takeaways

- Proof by contradiction: assume the opposite, derive an impossibility, conclude the original claim.
- √2 is irrational — proved by assuming it's rational, then showing that forces a contradiction.
- Proofs can lean on separately proved lemmas to stay organized.
- Circular reasoning — assuming the conclusion — invalidates an entire argument.

Reflection Questions

31. Why does the √2 proof need "lowest terms" at the start? What breaks without it?
32. The lemma was proved by showing "if p is odd, p^2 is odd" instead of directly proving "if p^2 is even, p is even." Why are these two statements logically the same thing?
33. Could you adapt this proof to show √3 is irrational? What would need to change?

Action Steps

34. From memory, with the book closed, write the full √2 proof. Check every step afterward.

35. Prove √3 is irrational. You'll need a lemma: if p^2 is divisible by 3, so is p. Prove that first.
36. Prove by contradiction that there are infinitely many prime numbers. (Assume there are finitely many, p_1 through p_n; consider $N = p_1 \times p_2 \times \ldots \times p_n + 1$.)

Chapter 7: Why Negative × Negative = Positive — A Rigorous Proof

> *"In mathematics, you don't understand things. You just get used to them." — John von Neumann (we respectfully disagree, at least for this chapter)*

The Question Everyone Has

Ask a room of fifteen-year-olds why a negative times a negative is positive, and most will either just repeat the rule or reach for an analogy: "the enemy of my enemy is my friend," or "two wrongs make a right." Almost nobody can actually derive it. This chapter does that — no analogies, just a clean logical chain from the properties in Chapter 5.

What We Will Prove

Two steps:

```
Step 1:  a × (−b) = −(ab)     [positive × negative = negative]
Step 2:  (−a) × (−b) = ab     [negative × negative = positive]
```

Everything else about signs follows immediately from these two.

Step 1: Positive × Negative = Negative

Theorem: for all a, b in Z, $a \times (-b) = -(ab)$.

Proof. We know $b + (-b) = 0$ (Property 3). Multiply both sides by a:

```
a × (b + (-b)) = a × 0 = 0
```

Apply distributivity to the left side:

```
a×b + a×(-b) = 0
```

This says $a\times(-b)$ is the additive inverse of ab. Since additive inverses are unique, $a\times(-b) = -(ab)$. ∎

Check it: $3\times(-4) = -(3\times4) = -12$. ✓

Step 2: Negative × Negative = Positive

Theorem: for all a, b in Z, $(-a) \times (-b) = ab$.

Proof. Again start from $b + (-b) = 0$, but this time multiply both sides by $(-a)$:

```
(-a) × (b + (-b)) = (-a) × 0 = 0
```

Distribute:

$$(-a)\times b + (-a)\times(-b) = 0$$

From Step 1, $(-a)\times b = -(ab)$. Substitute:

$$-(ab) + (-a)\times(-b) = 0$$

So $(-a)\times(-b)$ is the additive inverse of $-(ab)$ — which is just ab (the inverse of a negative is the original positive). So:

$$(-a) \times (-b) = ab \quad \blacksquare$$

Check it: $(-3)\times(-4) = 3\times4 = 12$. ✓

> KEY INSIGHT
>
> The sign rules aren't arbitrary conventions someone made up — they're the *only* way to extend multiplication to negative numbers without breaking distributivity or additive inverses, both of which are non-negotiable. The rule is forced on us by the rest of the system.

Why Analogies Aren't Enough

Sign Rules for Multiplication

	× negative	× positive
positive ×	negative	positive
negative ×	positive	negative

Same sign → positive result. Different signs → negative result.

Proved rigorously in Chapter 7 — not a convention, a consequence.

The full sign-rule grid: same signs give positive, different signs give negative.

"Two wrongs make a right" is a fine memory hook, but it isn't a proof — it shows the rule is plausible, not that it's *necessary*. The derivation above shows that if you accept the nine properties from Chapter 5, you have no choice about the sign rules. They're forced.

The Sign Rules in the Real World

REAL EXAMPLE

> Think about a car in reverse gear (negative direction), backing up (negative displacement over time). The net motion — where the car ends up relative to where it started — comes out positive, in the direction it started facing. Negative direction combined with negative displacement gives a positive result, exactly matching the sign rule.

> REAL EXAMPLE
>
> Or think about debt shrinking over time: if your loan balance is decreasing (a negative rate of change) and you look backward in time (a negative time direction), the math correctly tells you your balance was *higher* in the past — a positive difference — which matches reality.

Computers rely on this too: the two's-complement system used to represent negative numbers in virtually every processor is designed specifically so ordinary binary multiplication automatically produces the correct sign, every time, without special-casing negatives.

Common Sign Errors — and Why Understanding Fixes Them

- **$-(a+b) = -a + b$** is wrong. Correctly, $-(a+b) = -a - b$ — the negative distributes over *both* terms.
- **$(-a)^2 = -a^2$** is wrong. Correctly, $(-a)^2 = (-a)(-a) = a^2$ — squaring a negative always gives a positive.
- **$\sqrt{(a^2)} = a$** is wrong in general. Correctly, $\sqrt{(a^2)} = |a|$ — if a is negative, $\sqrt{(a^2)} = -a$, not a.

Each of these mistakes comes from misapplying the sign rules. Once you understand *why* negative times negative is positive — not as a memorized rule but as a forced consequence of the axioms — these errors stop happening, because you can rebuild the correct answer from scratch every time.

Key Takeaways

- Positive × negative = negative: proved from distributivity and the additive inverse.
- Negative × negative = positive: proved the same way, one layer deeper.
- The sign rules are forced by the axioms, not chosen by convention.
- Analogies describe; only the axiomatic proof actually explains.

Reflection Questions

37. In your own words, with no formulas, explain why negative times negative has to be positive.
38. Both proofs lean on distributivity at the key step. What would break if integers didn't have that property?
39. Both proofs use *uniqueness* of the additive inverse. Why does uniqueness matter here?

Action Steps

40. Write both proofs from memory, then check them against the chapter.
41. Compute $(-3)(-4)(-2)$, showing every step. State a general rule for predicting the sign of a product from how many negative factors it has.
42. Prove that $(-a)(-b)(-c) = -(abc)$ for any integers a, b, c, using only this chapter's results.

Chapter 8: Solving Integer Equations — An Elegant Technique

> *"A mathematician, like a painter or a poet, is a maker of patterns." — G. H. Hardy*

A Different Kind of Constraint

Ordinary algebra looks for all real-number solutions to an equation. Integer equations add an extra rule: solutions must be whole numbers. That one constraint changes everything. The equation $xy = x + y$ has infinitely many real-number solutions, but exactly two integer ones — and there's a systematic way to find them all.

These are called **Diophantine equations**, after Diophantus of Alexandria, who studied integer solutions to equations back in the 3rd century CE. It's still an active area of number theory today — Fermat's Last Theorem, which took over 350 years to finally prove, is a Diophantine equation.

The Core Technique: Factor, Then Enumerate

The method has three moves: **rearrange** (get everything on one side, a constant on the other), **factor** (add a carefully chosen number to both sides so the left side splits into two single-variable expressions multiplied together), and **enumerate** (list every integer pair whose product matches the right side, then solve each case). Since the right side is a fixed number, it only has finitely many factor pairs — you're guaranteed to find every solution.

> REAL EXAMPLE
>
> Think of a small theater selling adult tickets at one price and child tickets at another, and you know total tickets sold and total revenue. That's a Diophantine setup — you need whole numbers of tickets, not fractions of a ticket, and this factoring technique is exactly how you'd find every combination that fits the numbers.

The General Formula

Any equation of the form $xy + \alpha x + \beta y = N$ can be factored using:

$$(x + \beta)(y + \alpha) = xy + \alpha x + \beta y + \alpha\beta$$

So adding $\alpha\beta$ to both sides of the original equation turns the left side into a clean product:

$$(x + \beta)(y + \alpha) = N + \alpha\beta$$

Worked Example: xy = x + y

Rearranged: $xy - x - y = 0$. Here $\alpha = \beta = -1$, so $\alpha\beta = 1$. Add 1 to both sides:

$$(x - 1)(y - 1) = 1$$

The only integer pairs multiplying to 1 are (1,1) and (−1,−1):

- $x-1=1$, $y-1=1$ → $x=2$, $y=2$. Check: $2\times2 = 4 = 2+2$. ✓
- $x-1=-1$, $y-1=-1$ → $x=0$, $y=0$. Check: $0\times0 = 0 = 0+0$. ✓

Those are the only two integer solutions — exactly as promised at the start of the chapter.

Worked Example: xy − x + y = 3

Here $\alpha = -1$, $\beta = 1$, so $\alpha\beta = -1$. Formula: $(x+1)(y-1) = 3 + (-1) = 2$.

$$(x + 1)(y - 1) = 2$$

Integer factor pairs of 2: (1,2), (2,1), (−1,−2), (−2,−1). Each gives one solution:

- (1,2): $x=0$, $y=3$. Check: $0-0+3=3$. ✓
- (2,1): $x=1$, $y=2$. Check: $2-1+2=3$. ✓
- (−1,−2): $x=-2$, $y=-1$. Check: $2+2-1=3$. ✓
- (−2,−1): $x=-3$, $y=0$. Check: $0+3+0=3$. ✓

A Note on Prime Right-Hand Sides

If the number on the right side of your factored equation happens to be prime, you'll always get exactly four solution pairs — because a prime number only factors as (1, itself) or (−1, −itself), each in two orders. That's a useful shortcut for checking your work.

Key Takeaways

- Diophantine equations require whole-number solutions only — a constraint that turns infinite possibilities into a short, findable list.
- The method: rearrange, factor (add $\alpha\beta$ to both sides), then enumerate every integer factor pair.
- When the right-hand side is prime, expect exactly four solutions.

Reflection Questions

43. Why does restricting to integer solutions turn an infinite solution set into a finite one?
44. What happens to this method if N turns out to be zero? Work through an example.
45. Could this technique extend to three variables? What would change?

Action Steps

46. Solve $xy + 2x - y = 11$ for integers, showing the full factor-and-enumerate process.
47. Make up your own two-variable equation, solve it, then swap with a study partner (or check it yourself a day later) to verify.
48. Research one other famous Diophantine equation (Pell's equation is a good one) and summarize, in your own words, why it's interesting.

Chapter 9: Generalization — The Mathematician's Superpower

> "*Mathematics is the art of giving the same name to different things.*" — *Henri Poincaré*

From One Case to All Cases

The story goes that a young Carl Friedrich Gauss, asked by a teacher to add up the numbers 1 through 100 as busywork, produced the answer almost instantly. His trick: pair the first and last numbers (1+100=101), the second and second-to-last (2+99=101), and so on — 50 pairs, each summing to 101, giving 50×101 = 5,050.

That's a clever trick for one specific problem. What makes it *mathematics* rather than just a party trick is the next move: generalizing it. The same pairing works for 1 through n, for any n, giving the formula n(n+1)/2 — which we already proved by induction in Chapter 3. One clever observation about one number became a permanent tool for every number.

KEY INSIGHT

Generalization is the move from "here's a trick that works for this problem" to "here's why it must work for every problem of this shape." It's arguably the single most valuable habit of mind in mathematics.

Generalizing a Pattern

REAL EXAMPLE

Suppose a shop offers a fixed discount, then charges tax on the discounted price. You might compute one example carefully — a ₹1,000 item, 20% off, then 10% tax — and get ₹880. Generalizing means asking: does it matter whether the discount or the tax gets applied first? Working the algebra symbolically (price × 0.8 × 1.10 versus price × 1.10 × 0.8) shows the order doesn't matter, because multiplication commutes. One receipt turned into a permanent, reusable fact.

The Generalization Habit

Every time you solve a specific problem, ask: *what made this work, and where else does that same reason apply?* That question is what turns isolated answers into transferable understanding. It's also exactly what separates "I got this one problem right" from "I now understand this whole category of problems."

From Specific Proofs to General Theorems

Look back at the proofs in Chapters 6 and 7. The $\sqrt{2}$ proof wasn't really about the specific number 2 — the same argument, with minor changes, proves $\sqrt{3}$, $\sqrt{5}$, or $\sqrt{p}$ irrational for any prime p. The sign-rule proof in Chapter 7 wasn't really about any specific a and b — it worked for *every* pair of integers, because it never used a specific value, only the axioms.

That's the hallmark of a genuinely general proof: nowhere does it sneak in a fact that's true for the example but not for the general case.

Key Takeaways

- Generalization turns a specific trick into a permanent, reusable tool.

- Gauss's pairing trick for 1 through 100 generalizes to n(n+1)/2 for any n.
- Good general proofs never secretly rely on properties specific to one example.
- Asking "why did this work, and where else does that reason apply?" is a habit worth building deliberately.

Reflection Questions

49. Pick a math trick you know (a shortcut, a mental-math technique). Can you generalize it? Where does it stop working, and why?
50. Why does the √2 irrationality proof generalize so easily to other square roots?
51. What's the difference between "I checked this for five examples" and "I proved this for all cases"?

Action Steps

52. Adapt the √2 proof to show √5 is irrational. Note exactly which lines change and which stay identical.
53. Take Gauss's pairing trick and generalize it to the sum of the first n *even* numbers. Derive a formula and test it on a small case.
54. Find one "trick" you use in daily life (a budgeting shortcut, a cooking ratio) and write down the general rule underneath it.

Chapter 10: Even, Odd, and the Mathematics of Parity

> *"God made the integers; all else is the work of man." — Leopold Kronecker*

Defining Parity

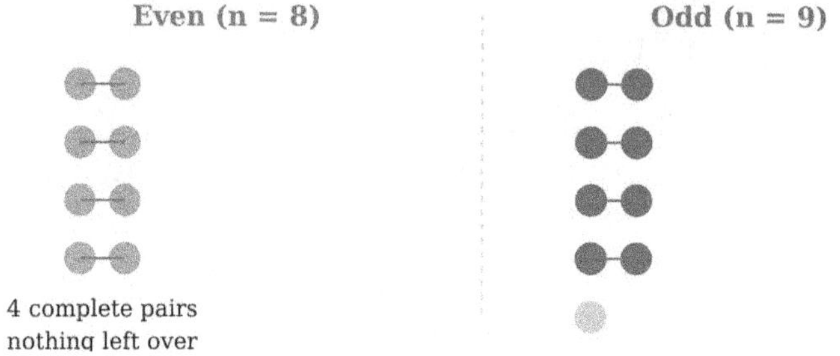

Even numbers pair up completely; odd numbers always leave exactly one unpaired.

An integer n is **even** if n = 2k for some integer k, and **odd** if n = 2k+1 for some integer k. Every integer is one or the other, never both, never neither — a fact so familiar it's easy to forget it's actually provable, not just observed.

REAL EXAMPLE

Think of pairing up shoes. An even number of shoes pairs up with nothing left over; an odd number always leaves exactly one shoe standing alone. That "leftover of exactly one, or none" is precisely what evenness and oddness mean.

The Arithmetic of Parity

A few quick, provable facts:

- Even + even = even. Odd + odd = even. Even + odd = odd.
- Even × anything = even.
- Odd × odd = odd.

Let's prove the first one properly, as practice with the tools from earlier chapters. Let $a = 2j$ and $b = 2k$ (both even). Then $a + b = 2j + 2k = 2(j+k)$, which is even by definition, since $j+k$ is an integer. That's it — a direct proof, no contradiction needed this time.

PLAIN ENGLISH

Not every proof needs contradiction. A *direct proof* just starts from the definitions and follows the logic straight through to the conclusion — often the cleanest option when the path from assumption to conclusion is short.

A Locker-Room Puzzle

Lockers 1-12 after all 12 students pass through

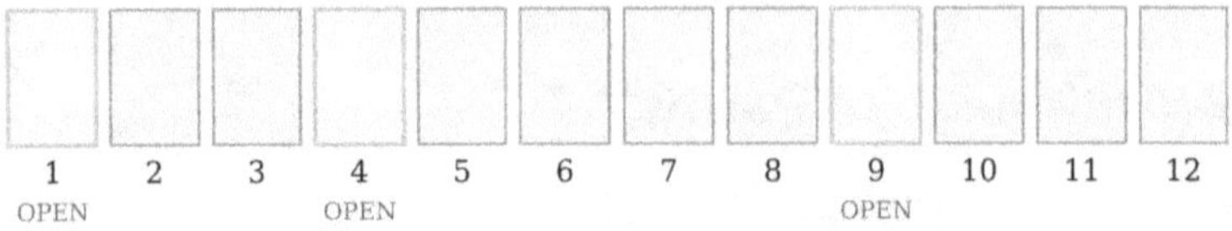

Open lockers: 1, 4, 9 — exactly the perfect squares

Only the perfect-square lockers end up open.

REAL EXAMPLE

Picture 100 lockers, all closed, and 100 students. Student 1 opens every locker. Student 2 toggles every 2nd locker (closes it, since it was open). Student 3 toggles every 3rd locker. This continues through student 100. Which lockers end up open?

A locker gets toggled once for every factor of its number — locker 12 gets toggled by students 1, 2, 3, 4, 6, and 12. A locker ends up open only if it's toggled an *odd* number of times, which happens only when its number of factors is odd. And a number has an odd number of factors only when it's a perfect square (because factors normally pair up, like 3 and 4 for 12, except when a factor pairs with itself, as 6 does with itself for 36). So the open lockers at the end are exactly 1, 4, 9, 16, 25, ... — the perfect squares. Parity, hiding inside a puzzle that looks like it has nothing to do with even and odd numbers.

The Parity Theorem: A Capstone Result

Theorem: if n^2 is even, then n is even.

Proof (by contrapositive): Suppose n is odd, so $n = 2k+1$. Then $n^2 = 4k^2 + 4k + 1 = 2(2k^2+2k) + 1$, which is odd. So an odd n always gives an odd n^2 — meaning an even n^2 can never have come from an odd n. Therefore, if n^2 is even, n must be even. ∎

This is exactly the lemma we leaned on in Chapter 6 to prove √2 is irrational — parity wasn't a side detail there, it was the engine driving the whole argument.

Why This Chapter Is the Capstone

Notice what just happened: this short chapter used direct proof, contrapositive, a real-world puzzle, and a callback to one of the book's central results — all built from a two-line definition. That's the whole promise of this book in miniature: start from something almost embarrassingly simple, reason carefully, and end up somewhere genuinely deep.

Key Takeaways

- Every integer is either even or odd, never both, never neither.
- Even+even=even, odd+odd=even, even+odd=odd; even×anything=even; odd×odd=odd.
- Not every proof needs contradiction — direct proof and contrapositive are often cleaner.
- "If n^2 is even, then n is even" is the parity theorem, and it's the engine behind the √2 proof from Chapter 6.

Reflection Questions

55. Prove that odd × odd = odd, using the same style of direct proof used for even + even.
56. Why does a locker number end up open exactly when it has an odd number of factors?
57. Compare the contrapositive proof of the parity theorem to the contradiction-based proof in Chapter 3. When would you reach for one over the other?

Action Steps

58. Prove that the square of an odd number is always odd, directly (not by contrapositive).
59. Work the locker puzzle for lockers 1 through 30 by hand, and confirm the perfect-square pattern yourself.
60. Look back at every proof in this book. For each one, note whether it used direct proof, contradiction, contrapositive, or induction.

What pattern do you notice about when each gets used?

Conclusion: Where the Road Leads

Go back to the sentence you wrote in Chapter 1, about your relationship with math. Read it again now.

You started this book with a handful of properties that looked almost too obvious to bother stating, and by the end you'd used them to prove that a negative times a negative is positive, that the square root of two can't be written as a fraction, that there's no biggest whole number, and that every locker with a perfect-square number ends up open. None of that was memorized. All of it was built, one justified step at a time, from nine simple properties and two proof techniques.

That's the actual point of this book. Not the specific theorems — useful as they are — but the demonstration that mathematical understanding is *available* to you. It was never about being a "math person." It was about whether the gaps got closed.

They're closed now. What you do with that is up to you — and the next chapter gives you a concrete path if you want to keep going.

Mathematical Roadmap: Your Next Steps

This isn't a reading list. It's a sequenced path from this book to real undergraduate-level mathematics.

Stage 1 — This Book (Finish It First)

Work every chapter exercise. Before moving on, you should be able to reconstruct these five results from memory: no largest natural number, √2 is irrational, negative × negative = positive, the parity theorem, and one proof of your choosing from Appendix A. Budget four to six weeks at a few sessions a week.

Stage 2 — Introduction to Proof Writing

Recommended text: **How to Prove It** by Daniel Velleman.

This book develops the full toolkit of proof techniques — direct proof, contrapositive, induction, and more — and formally introduces sets and functions. It bridges the gap between this book and university-level mathematics.

Stage 3 — Number Theory

Recommended text: **Elementary Number Theory** by David Burton.

The natural next step: primes, divisibility, and modular arithmetic, deepening the integer arithmetic from Chapters 5, 6, and 10. You'll meet theorems by Fermat, Euler, and Gauss — many proved with techniques you already know.

Stage 4 — Abstract Algebra

Recommended text: **A First Course in Abstract Algebra** by John Fraleigh.

This generalizes the nine properties from Chapter 5 into the study of groups, rings, and fields — abstract structures defined purely by which axioms they satisfy. The integers form a ring; the rationals form a field. This is where you see *why* the properties in Chapter 5 take exactly the form they do.

Stage 5 — Real Analysis

Recommended text: **Understanding Analysis** by Stephen Abbott.

This is where calculus gets its rigorous foundations — limits, continuity, differentiation, integration, all proved from first principles using the same kind of reasoning developed in this book.

A Note on Pace

The full path from here to real analysis typically takes three to five years of consistent, unhurried study. That's not a discouragement — it's an honest calibration. Mathematics rewards patience above every other virtue, and each stage is a satisfying destination in its own right. You don't need to reach Stage 5 for this to have been worth it. Stage 1 alone, done properly, changes how you think about every quantitative thing you'll ever encounter.

The road ahead is long. It's also genuinely beautiful.

Begin.

Appendix E: Practice Problems with Full Solutions

Ten problems, spanning every chapter. Attempt each one fully before reading the solution — the struggle is where the learning happens.

Problem 1

Prove that the sum of any two odd integers is even.

> SOLUTION
>
> Let the two odd integers be $2j+1$ and $2k+1$ for integers j, k. Their sum is $2j+1+2k+1 = 2j+2k+2 = 2(j+k+1)$. Since $j+k+1$ is an integer, the sum is even by definition. ∎

Problem 2

Solve for integers: $xy + 3x - 2y = 1$.

> SOLUTION
>
> Here $\alpha = 3$, $\beta = -2$, so $\alpha\beta = -6$. Add -6 to both sides: $(x-2)(y+3) = 1 - 6 = -5$. The integer factor pairs of -5 are $(1,-5)$, $(-1,5)$, $(5,-1)$, $(-5,1)$. Working each back through $x-2$ and $y+3$ gives four solutions: $(x,y) = (3,-8)$, $(1,2)$, $(7,-4)$, $(-3,0)$. Check any one: for $(1,2)$, $1\times2+3\times1-2\times2$

$= 2+3-4 = 1.$ ✓

Problem 3

Prove that if a is even and b is any integer, then ab is even.

SOLUTION

Since a is even, $a = 2k$ for some integer k. Then $ab = (2k)b = 2(kb)$, which is even by definition, since kb is an integer regardless of what b is. ∎

Problem 4

Using the properties from Chapter 5, prove that $-(-a) = a$ for any integer a.

SOLUTION

By definition, $-a$ is the number that satisfies $a + (-a) = 0$. By the same definition, $-(-a)$ is the number that satisfies $(-a) + (-(-a)) = 0$. But a already satisfies $(-a) + a = 0$ (commuting Property 3). Since additive inverses are unique, $-(-a)$ must equal a. ∎

Problem 5

Compute, showing every step: $(-7) \times 4 + (-7) \times (-4)$.

SOLUTION

By distributivity, $(-7)\times4 + (-7)\times(-4) = (-7)\times(4+(-4)) = (-7)\times0 = 0$. (Direct computation confirms it: $-28 + 28 = 0$.)

Problem 6

Prove by contradiction: there is no smallest positive rational number.

SOLUTION

Suppose q were the smallest positive rational. Then $q/2$ is also a positive rational (half of a ratio of integers is still a ratio of integers), and $q/2 < q$. This contradicts q being smallest. So no smallest positive rational exists. ∎

Problem 7

Is the set of even integers closed under addition? Under multiplication? Prove both answers.

SOLUTION

Yes to both. For addition: $2j + 2k = 2(j+k)$, even. For multiplication: $(2j)(2k) = 4jk = 2(2jk)$, even. Both sums and products of even integers stay even, so the set is closed under both operations.

Problem 8

A father is three times as old as his son. In 12 years, he will be twice as old. How old is each now? (Set up and solve as an integer equation.)

SOLUTION

Let the son's age be x, so the father is $3x$. In 12 years: $3x+12 = 2(x+12)$. Expand: $3x+12 = 2x+24$, so $x=12$. The son is 12, the father is 36. Check: in 12 years, 24 and 48 — and 48 is indeed twice 24. ✓

Problem 9

Prove that the square of any integer is either divisible by 4 (if the integer is even) or leaves remainder 1 when divided by 4 (if odd).

SOLUTION

If $n=2k$, then $n^2 = 4k^2$, clearly divisible by 4. If $n=2k+1$, then $n^2 = 4k^2+4k+1 = 4(k^2+k)+1$, which leaves remainder 1 when divided by 4. Since every integer is even or odd, one of these two cases always applies. ∎

Problem 10

Find all integer solutions to $x^2 - y^2 = 24$. (Hint: factor as $(x-y)(x+y) = 24$, and remember $x-y$ and $x+y$ must have the same parity.)

SOLUTION

Since $(x-y)(x+y)=24$ and $x-y$, $x+y$ always share the same parity (both even or both odd — check: their sum is $2x$, always even), and 24 is even, both factors must be even. The even factor pairs of 24 are (2,12), (12,2), (4,6), (6,4), and their negatives. Working through $(x-y,x+y)=(2,12)$ gives $x=7$, $y=5$. Working through (4,6) gives $x=5$, $y=1$. Together with sign variations, the full solution set is $(x,y) = (\pm7,\pm5)$ and $(\pm5,\pm1)$, with matching signs on each pair. Check one: $7^2-5^2 = 49-25 = 24$. ✓

Appendix F: Famous Mathematical Proofs

A short tour of results outside this book's main path, chosen because each one uses the exact same tools you've now practiced: direct proof, contradiction, and induction.

The Infinitude of Primes (Euclid, c. 300 BCE)

Theorem: there are infinitely many prime numbers.

Proof (by contradiction). Suppose there are only finitely many primes: p_1, p_2, ..., p_n. Consider the number $N = p_1 \times p_2 \times \ldots \times p_n + 1$. N is either prime itself, or divisible by some prime. But N leaves remainder 1 when divided by any of p_1 through p_n (since each divides the product exactly), so none of our "complete" list of primes divides N. That means N's prime factors — and N must have at least one, being greater than 1 — are not on our list. This contradicts the assumption that p_1 through p_n were *all* the primes. So there must be infinitely many. ∎

KEY INSIGHT

Notice the shape: exactly the same contradiction structure as "no largest natural number" from Chapter 3. Assume a finite/bounded collection, construct something that must escape it, contradiction.

The Pythagorean Theorem

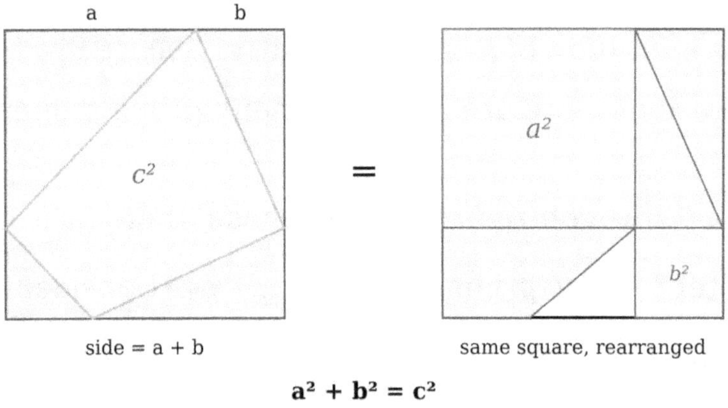

Two ways of computing the same square's area — the rearrangement that proves $a^2+b^2=c^2$.

Theorem: in a right triangle with legs a and b and hypotenuse c, $a^2 + b^2 = c^2$.

Proof (one of hundreds known). Take four copies of the right triangle and arrange them inside a square of side (a+b), forming a smaller tilted square of side c in the middle. The big square's area is $(a+b)^2$. That area equals the four triangles (each with area ½ab) plus the inner square (area c^2):

$$(a+b)^2 = 4(\tfrac{1}{2}ab) + c^2$$
$$a^2 + 2ab + b^2 = 2ab + c^2$$
$$a^2 + b^2 = c^2 \quad ∎$$

Sum of the First n Squares

Theorem: $1^2 + 2^2 + \ldots + n^2 = n(n+1)(2n+1)/6$.

Proof (by induction). Base case (n=1): $1 = 1(2)(3)/6 = 1$. ✓

Inductive step: assume the formula holds for n. Then:

$$1^2+2^2+\ldots+n^2+(n+1)^2 = n(n+1)(2n+1)/6 + (n+1)^2$$

Factor out (n+1)/6:

$$= (n+1)[n(2n+1) + 6(n+1)] / 6 = (n+1)(2n^2+7n+6)/6 = (n+1)(n+2)(2n+3)/6$$

This is exactly the formula with n replaced by n+1. By induction, the formula holds for all natural numbers. ∎

The Arithmetic Mean–Geometric Mean Inequality (Two-Variable Case)

Theorem: for any two non-negative real numbers a and b, $(a+b)/2 \geq \sqrt{ab}$.

Proof (direct). Start from a fact that's always true: $(\sqrt{a} - \sqrt{b})^2 \geq 0$, since any real number squared is non-negative. Expand:

$$a - 2\sqrt{ab} + b \geq 0$$
$$a + b \geq 2\sqrt{ab}$$
$$(a+b)/2 \geq \sqrt{ab} \quad \blacksquare$$

Equality holds exactly when a = b — when the square we started from is zero.

> REAL EXAMPLE
>
> This inequality is why, for a fixed perimeter, a square encloses more area than any other rectangle: it's the same statement as "arithmetic mean ≥ geometric mean," applied to the two

side lengths.

Appendix G: A Brief History of the Mathematics in This Book

Brahmagupta (598–668 CE)

The Indian mathematician and astronomer Brahmagupta gave the first systematic rules for arithmetic with zero and negative numbers in his 628 CE treatise, the *Brahmasphutasiddhanta*. He described negative numbers as "debts" and positive numbers as "fortunes" — a framing that still shows up in how negative numbers get taught today. He got most of his rules for zero right; his rule for dividing by zero was wrong, and it would take mathematicians centuries more to work out why that operation has no answer at all (Chapter 4).

Diophantus of Alexandria (c. 200–284 CE)

Diophantus wrote the *Arithmetica*, a collection of problems seeking whole-number or rational solutions to equations — the origin of the term "Diophantine equation" used in Chapter 8.

Remarkably little is known about his life; even his age at death is inferred from a puzzle written about him, which is itself a Diophantine equation.

René Descartes (1596–1650) and the Resistance to Negative Numbers

Descartes, one of the founders of modern algebra, still referred to negative solutions of equations as "false roots" — evidence of how long it took even brilliant mathematicians to fully accept negative numbers as legitimate (Chapter 4). Full acceptance didn't arrive until the 18th and 19th centuries, alongside a more rigorous, axiom-based treatment of number systems.

Carl Friedrich Gauss (1777–1855)

Gauss, one of history's greatest mathematicians, is said to have astonished his teacher at around age eight by instantly summing the numbers 1 through 100 using the pairing trick shown in Chapter 9. Whether or not the classroom story is exactly accurate, the technique and the formula n(n+1)/2 are real, and the underlying idea — pair the extremes, notice the pattern, generalize — is a genuine hallmark of how Gauss worked throughout his career.

Giuseppe Peano (1858–1932)

The Italian mathematician Peano formalized the natural numbers with a small set of axioms in 1889, giving arithmetic the same kind of rigorous foundation that geometry had had since Euclid. The Peano axioms — essentially, that there's a starting number and a well-behaved "successor" operation — sit quietly underneath the informal definition of N used in Chapter 3.

Georg Cantor (1845–1918)

Cantor's work on the size of infinite sets (Chapter 3) was so radical that it drew open hostility from some of his most senior contemporaries — the influential mathematician Leopold Kronecker reportedly called him a "corrupter of youth." Cantor experienced serious depressive episodes later in life; historians still debate how much was linked to this professional conflict. His set theory is now considered one of the foundational achievements of modern mathematics, taught to every undergraduate.

Appendix H: How to Study Mathematics — A Practical Guide

Work every problem before checking the answer. Reading a solution and understanding it feels like learning; it usually isn't. Struggling with a problem for ten minutes, even unsuccessfully, builds more than reading five solved examples.

Explain it out loud. If you can't explain a proof or definition to an imaginary student without notes, you don't fully own it yet. This is the fastest way to find the exact spot where your understanding is thin.

Revisit, don't just move on. Come back to Chapter 3's proofs a week after finishing Chapter 5. Understanding that survives a week without review is understanding that's actually yours.

Get comfortable being stuck. Being stuck is not a signal to quit — it's the state you're in right before a real idea lands. Treat the discomfort as information, not as a verdict on your ability.

Separate "I followed it" from "I could rebuild it." Following someone else's proof is easy. Reconstructing it from a blank page a day later is the real test — and it's the only kind of understanding that transfers to a new, unfamiliar problem.

Keep a running list of your "why" questions. Every time a rule feels arbitrary, write the question down, even if you can't answer it yet. Half of them will resolve themselves as you keep reading; the other half are exactly what to go research.

Appendix I: Proof of Key Results — A Compiled Reference

Every major theorem from the main chapters, restated in one place for quick review.

N has no largest element (Chapter 3). Assume M is the largest natural number. M+1 is also natural and exceeds M — contradiction. So no largest element exists.

√2 is irrational (Chapter 6). Assume $\sqrt{2} = p/q$ in lowest terms. Then $p^2 = 2q^2$, forcing p even; writing $p=2m$ gives $q^2 = 2m^2$, forcing q even too — contradicting "lowest terms." So no such fraction exists.

a × (−b) = −(ab) (Chapter 7). From $b + (-b) = 0$, multiply by a and distribute: $ab + a(-b) = 0$, so $a(-b)$ is the additive inverse of ab, i.e., $-(ab)$.

$(-a) \times (-b) = ab$ (Chapter 7). From $b + (-b) = 0$, multiply by $(-a)$ and distribute: $(-a)b + (-a)(-b) = 0$. Since $(-a)b = -(ab)$ from the previous result, substituting and solving gives $(-a)(-b) = ab$.

If n^2 is even, then n is even (Chapter 10, proved by contrapositive). If n is odd, $n=2k+1$, so $n^2 = 4k^2+4k+1$ is odd. So an even n^2 can only come from an even n.

$a \times 0 = 0$ (Chapter 5). $a\times0 = a\times(0+0) = a\times0 + a\times0$ by distributivity. Subtracting $a\times0$ from both sides gives $0 = a\times0$.

There are infinitely many primes (Appendix F, Euclid). Assuming a finite complete list $p_1,\ldots,p_n$, the number $p_1\times\ldots\times p_n+1$ has a prime factor not on the list — contradiction.

The Pythagorean theorem: $a^2+b^2=c^2$ (Appendix F). Proved by comparing two ways of computing the area of a square built from four copies of a right triangle plus a central tilted square.

Keep this page as a quick-reference sheet — if you can reconstruct every proof above from just its one-line statement, you've internalized the core of this book.

Appendix J: Extended Problem Set — Ten More Problems

These are a step harder than Appendix E. Take your time.

Problem 11

Prove that between any two distinct integers a and b, there are only finitely many integers, but between any two distinct rational numbers there are infinitely many rationals.

> SOLUTION
> If $a < b$ are integers, the integers strictly between them are $a+1$, $a+2$, ..., $b-1$ — a finite list of exactly $b-a-1$ numbers. But if $p/q < r/s$ are rationals, their average $(p/q + r/s)/2$ is a rational strictly between them; repeating this on the new, narrower interval produces another rational between the two, forever. So there are infinitely many rationals between any two distinct rationals, but only finitely many integers between any two distinct integers.

Problem 12

Find all integer solutions to $2xy - 4x + y = 5$.

SOLUTION

Divide the whole equation by common structure carefully: $2xy - 4x + y = 5$ doesn't factor as cleanly as our α/β formula (which assumes a coefficient of 1 on xy), so first divide by 2 conceptually: $xy - 2x + y/2 = 5/2$, which isn't integer-friendly. Instead, rearrange directly: $x(2y-4) = 5-y$, so $x = (5-y)/(2y-4)$, valid only when $(2y-4)$ divides $(5-y)$ evenly. Testing small integer values of y: y=3 gives $x=(5-3)/(6-4)=2/2=1$ — a valid solution $(x,y)=(1,3)$. y=1 gives $x=(5-1)/(2-4)=4/(-2)=-2$ — valid: $(-2,1)$. y=5 gives $x=0/6=0$ — valid: $(0,5)$. y=7 gives $x=(-2)/10$, not an integer. Checking further values shows no more integer solutions turn up as $|y|$ grows, since the fraction stops landing on whole numbers. Final solutions: (1,3), (−2,1), (0,5).

Problem 13

Prove that √6 is irrational, adapting the √2 proof from Chapter 6.

SOLUTION

Assume $\sqrt{6} = p/q$ in lowest terms. Squaring: $p^2 = 6q^2$. So p^2 is divisible by 6, meaning p^2 is divisible by both 2 and 3, so p is divisible by both 2 and 3 (using the same "if p^2 divisible by a prime, so is p" lemma style as Chapter 6), so p is divisible by 6; write $p=6m$. Then $36m^2 = 6q^2$, so $q^2 = 6m^2$, meaning q is also divisible by 6 — contradicting "lowest terms." So $\sqrt{6}$ is irrational.

Problem 14

A rectangle has integer side lengths and a perimeter of 24. List every possible integer area.

SOLUTION

Perimeter 24 means length + width = 12. Integer pairs (l,w) with $l+w=12$, $l\geq w\geq 1$: (11,1), (10,2), (9,3), (8,4), (7,5), (6,6). Their areas: 11, 20, 27, 32, 35, 36. Notice the area increases as the rectangle gets closer to a square — a concrete instance of the AM-GM inequality from Appendix F.

Problem 15

Prove: the product of any three consecutive integers is divisible by 6.

SOLUTION

Among any three consecutive integers, at least one is even (parity alternates), so the product is divisible by 2. Also, among any three consecutive integers, exactly one is divisible by 3 (since remainders mod 3 cycle through 0, 1, 2 every three numbers). So the product is divisible by both 2 and 3, hence by 6.

Problem 16

Solve for integers: $x^2 - 5x + y^2 - 3y = 0$, if you also know x and y are both positive. (Hint: complete the square in both variables.)

SOLUTION

Complete the square: $(x-5/2)^2 - 25/4 + (y-3/2)^2 - 9/4 = 0$, so $(x-5/2)^2 + (y-3/2)^2 = 34/4 = 8.5$. Multiply through by 4: $(2x-5)^2 + (2y-3)^2 = 34$. Now search for integer squares summing to 34: 25+9=34 works (5^2 and 3^2), and that's the only way with both terms being perfect squares under 34. So $2x-5=\pm5$ and $2y-3=\pm3$. Taking positive x,y: $2x-5=5$ gives $x=5$; $2y-3=3$ gives $y=3$. Check: $25-25+9-9=0$. ✓ So $(x,y)=(5,3)$.

Problem 17

Prove that if n is a positive integer, $n^2 + n$ is always even.

SOLUTION

$n^2 + n = n(n+1)$, the product of two consecutive integers. One of any two consecutive integers is always even, so their product is always even (Problem 3's result: even × anything = even).

Problem 18

Show that 0.999... (repeating) equals exactly 1, not "almost" 1.

SOLUTION

Let x = 0.999... Then 10x = 9.999... Subtracting: 10x − x = 9.999... − 0.999... = 9, so 9x = 9, so x = 1. There's no room for "almost" in this equation — the algebra forces exact equality.

Problem 19

Prove that the sum of the first n odd numbers (1+3+5+...) equals n^2.

SOLUTION

By induction. Base case n=1: $1 = 1^2$. Inductive step: assume $1+3+...+(2n-1) = n^2$. The next odd number is 2n+1, so $1+3+...+(2n-1)+(2n+1) = n^2 + 2n + 1 = (n+1)^2$. This

> matches the formula for n+1, completing the induction.

Problem 20

A puzzle: three friends split a restaurant bill. Together they pay ₹300, splitting it unevenly with only whole-rupee amounts, and no one pays the same amount as another. The largest share is exactly double the smallest. How many ways can this happen? (Set up as an integer equation and enumerate.)

> SOLUTION
>
> Let the shares be $a < b < c$ with $a+b+c=300$ and $c=2a$. Substituting: $a+b+2a=300$, so $b=300-3a$. We need $a < b < c$, i.e., $a < 300-3a < 2a$. From $a < 300-3a$: $4a<300$, so $a<75$. From $300-3a<2a$: $300<5a$, so $a>60$. So a ranges over integers 61 through 74 — 14 possible values, each giving one valid split $(a, 300-3a, 2a)$. For example, $a=61$ gives (61, 117, 122); check: $61+117+122=300$ and $122=2\times61$. ✓

Appendix K: Common Misconceptions and How to Fix Them

Every one of these mistakes was made by understanding a shortcut instead of the underlying idea. Each fix points back to the chapter that resolves it properly.

"Multiplying always makes numbers bigger."

True for two numbers greater than 1. False the moment a fraction or a negative number enters the picture: $5 \times 0.5 = 2.5$ (smaller), and $5 \times (-1) = -5$ (smaller still, and negative). The rule people actually mean is "multiplying by something greater than 1 increases magnitude" — see Chapter 7 for exactly how sign interacts with magnitude.

"A negative number is just a positive number with a minus sign stuck on."

This treats the minus sign as decoration rather than as information. Chapter 4 defines negative numbers properly: $-a$ is *the specific number* that satisfies a +

$(-a) = 0$. That's a functional definition, not a cosmetic one, and it's what makes the sign rules in Chapter 7 provable rather than just memorized.

"Dividing by zero gives infinity."

Calculators sometimes display "Error" or "∞" for this, which reinforces the myth. Chapter 4 shows the real reason it's undefined: $a \div 0 = c$ would require $0 \times c = a$, which is either impossible (if $a \neq 0$) or true for every c (if $a = 0$) — neither case gives a single well-defined answer. "Infinity" is not a number that solves this; it's a description of unbounded growth, a different concept entirely (see Chapter 3).

"$\sqrt{(a^2)}$ always equals a."

Only true when a is non-negative. If $a = -3$, then $a^2 = 9$, and $\sqrt{9} = 3$, not -3. The correct general statement is $\sqrt{(a^2)} = |a|$. This misconception directly causes sign errors in algebra — see Chapter 7's "Common Sign Errors" section.

"A mathematical proof is just a very convincing explanation."

A proof is not about persuasiveness — a wrong argument can be extremely persuasive, and a correct one can feel unintuitive. A proof is a chain of statements where each link is justified by a definition, an axiom, or an earlier proven result, with zero gaps. Chapter 6 walks through exactly what separates a real proof from a convincing-sounding argument, most importantly through the concept of circular reasoning.

"If it works for a few examples, it's true in general."

Checking examples builds intuition but never proves anything about the infinitely many cases you didn't check. Chapter 9 is built entirely around the gap between "I verified this for several cases" and "I proved this for all cases" — and Chapter 3's induction technique is precisely the tool that closes that gap rigorously.

"0 isn't really a number, it's just 'nothing.'"

Zero has real, provable arithmetic behavior: it's the unique additive identity ($a+0=a$ for every a) and it forces any product to be zero ($a \times 0=0$, proven directly from distributivity in Chapter 5). "Nothing" doesn't have algebraic properties. Zero does — that's exactly what makes it a number rather than an absence.

"Odd and even are just about the last digit."

Checking the last digit is a valid shortcut for base-ten numbers, but it's a consequence of the real definition, not the definition itself: n is even precisely when $n=2k$ for some integer k (Chapter 10). Understanding it that way is what lets you prove things about parity — like the locker puzzle or the parity theorem — rather than just sorting numbers into two piles.

Appendix L: Worked-Example Companion — One Extended Application per Chapter

Each entry below takes a chapter's core idea and works it through a single real situation in full detail, start to finish — useful as a template for building your own examples.

Chapter 1 — Foundations: Learning a Recipe vs. Understanding Cooking

A recipe tells you: sauté onions 5 minutes, add garlic 30 seconds, add tomatoes, simmer 20 minutes. Follow it exactly and you get a good sauce. But you're following, not understanding — change one ingredient (say, swap tomatoes for a squash) and the recipe stops working, because you never learned *why* those times and that order were chosen. Understanding cooking means knowing that onions need longer because raw starch takes time to soften and sweeten, that garlic burns fast and turns bitter past 30 seconds, and that simmering (not boiling) preserves texture while concentrating flavor. Once you know the *why,* you can adapt to a new ingredient on the fly. That gap — between following steps and understanding the principle behind them — is exactly the gap this book closes for arithmetic.

Chapter 2 — Counting: Splitting a Bill Without Numbers

Imagine four friends at dinner trying to split a bill fairly with no numbers allowed — only physical tokens (pebbles, say) representing rupees. Every payment, every share, every "you owe me" would require carrying and comparing piles of pebbles. The moment someone invents a symbol for "50 pebbles" that doesn't require actually holding 50 pebbles, bill-splitting stops being a physical logistics problem and becomes a symbolic one — exactly the leap from tally marks to abstract numerals traced in Chapter 2.

Chapter 3 — Natural Numbers and Infinity: A Savings App That Never Caps Out

Picture a savings app's balance display. No matter how much money accumulates, the app never needs to say "sorry, no more room for digits" — the display just grows another character when needed. That's the natural numbers' "no largest element" property, made concrete: the *system* itself places no ceiling on how large a count can go, even though any specific balance at any specific moment is, of course, finite.

Chapter 4 — Integers: Reading a Bank Statement

A real bank statement mixes credits (deposits, positive) and debits (withdrawals, negative) into one running balance. A statement showing −₹2,400 after a series of transactions isn't an error or a "meaningless" result — it's integers doing exactly what they were built for in Chapter 4: representing a deficit as a first-class number, with all the same arithmetic rules as a surplus.

Chapter 5 — Properties of Arithmetic: Rearranging a Moving Truck

Loading boxes into a moving truck in the order living-room, then kitchen, then bedroom uses the same truck space as kitchen, then bedroom, then living-room — the total volume doesn't care about loading order. That's associativity and commutativity of addition, made physical: (living + kitchen) + bedroom = living + (kitchen + bedroom), and the grouping and order are both free to change without changing the outcome.

Chapter 6 — Proof by Contradiction: A Detective's Alibi

A detective claims a suspect couldn't have committed a crime at 9 PM because three independent witnesses place them at a restaurant at that time. The reasoning: assume the suspect *was* at the crime scene at 9 PM; then they couldn't simultaneously be at the restaurant; but three witnesses say they were; contradiction; so the assumption is false. This is proof by contradiction doing real investigative work, structurally identical to the √2 proof in Chapter 6 — assume the claim you doubt, derive an impossibility, reject the assumption.

Chapter 7 — Sign Rules: A Video Editor Scrubbing Backward

In video-editing software, dragging the playhead backward through a clip that's itself playing in reverse produces forward-looking motion on screen — two "negatives" (backward scrubbing, reversed clip) combining into a "positive" (forward-seeming motion). It's a slightly playful illustration, but the underlying math is the same sign rule proved rigorously in Chapter 7: negative direction combined with negative direction yields positive.

Chapter 8 — Diophantine Equations: Packing Boxes of Two Sizes

A warehouse ships items in boxes of 5 units or 8 units only, and needs to fill an order for exactly 39 units using whole boxes. Setting up $5x + 8y = 39$ and searching integer solutions (x,y must be non-negative here) gives x=3, y=3 (15+24=39) as the only valid combination — a direct, practical instance of the integer-equation methods from Chapter 8, restricted to non-negative solutions because you can't ship a negative number of boxes.

Chapter 9 — Generalization: A Tip Calculator

Compute a 15% tip on a ₹840 bill: ₹126. That's one answer to one problem. Generalizing asks: does the *order* of computing tip-then-tax versus tax-then-tip change the final total? Writing it symbolically (bill × 1.15 × 1.18 vs. bill × 1.18 × 1.15) shows they're identical, by the same commutativity of multiplication proved in Chapter 5 — one receipt's arithmetic becomes a permanent fact about every receipt.

Chapter 10 — Parity: Splitting Teams Fairly

A gym class of 23 students can't be split into two perfectly equal teams — 23 is odd, so any split leaves one team with one more player. This isn't a scheduling inconvenience to work around cleverly; it's a direct, unavoidable consequence of what "odd" means (Chapter 10). No clever pairing rule can make an odd number split evenly — the parity theorem guarantees it in advance, before anyone even tries.

Appendix M: Glossary of Terms

Axiom — a statement accepted without proof as a starting assumption.

Associativity — a property where grouping doesn't matter: (a+b)+c = a+(b+c).

Commutativity — a property where order doesn't matter: a+b = b+a.

Contrapositive — proving "if not-Q then not-P" instead of "if P then Q"; logically equivalent, sometimes easier.

Diophantine equation — an equation for which only whole-number solutions are sought.

Distributivity — the property a×(b+c) = a×b + a×c, linking addition and multiplication.

Function — a rule that assigns exactly one output to each input from a given set; not covered in depth in this book, but foundational to the mathematics that follows it (see Appendix O).

Integer — a member of Z = {..., −2, −1, 0, 1, 2, ...}: a whole number, positive, negative, or zero.

Irrational number — a real number that cannot be written as a fraction of two integers.

Natural number — a member of N = {1, 2, 3, ...}, the counting numbers.

Parity — whether a number is even or odd.

Proof — a chain of statements, each justified by a definition, an axiom, or an earlier proven result, that establishes a claim beyond doubt. See Chapter 6 for what separates a real proof from a merely convincing argument.

Proof by contradiction — assuming the opposite of a claim, showing it leads to an impossibility, and concluding the claim is true.

Rational number — any number expressible as a fraction p/q of integers, $q \neq 0$.

Set — a well-defined collection of distinct objects.

Theorem — a statement that has been proved true, using only axioms, definitions, and previously established theorems.

Trichotomy law — for any two integers a and b, exactly one of $a<b$, $a=b$, $a>b$ holds.

Appendix N: Mathematical Notation Guide

Symbol	Meaning
$\in$	"is an element of"
$\notin$	"is not an element of"
$\subset$	"is a (proper) subset of"
$\forall$	"for all"
$\exists$	"there exists"
N	the natural numbers, {1, 2, 3, ...}
Z	the integers, {..., −1, 0, 1, ...}
Q	the rational numbers
R	the real numbers
∎	end of proof (also written Q.E.D.)

Appendix O: Recommended Further Reading

- **How to Prove It**, Daniel Velleman — the natural next step after this book for proof-writing fundamentals.
- **Elementary Number Theory**, David Burton — deepens everything from Chapters 5, 6, and 10.
- **A First Course in Abstract Algebra**, John Fraleigh — generalizes the nine properties of Chapter 5.
- **Understanding Analysis**, Stephen Abbott — the rigorous foundations of calculus.
- **Journey Through Genius**, William Dunham — an accessible history of great theorems, in the same spirit as this book's historical asides.

Appendix P: References and Bibliography

Abbott, S. (2015). *Understanding Analysis* (2nd ed.). Springer.

Aristotle. *Physics,* Book III (on the potential and actual infinite).

Brahmagupta (628 CE). *Brahmasphutasiddhanta.*

Diophantus of Alexandria. *Arithmetica.*

Euclid. *Elements,* Book IX, Proposition 20 (infinitude of primes).

Fraleigh, J. B. (2002). *A First Course in Abstract Algebra* (7th ed.). Pearson.

Galilei, G. (1638). *Discorsi e Dimostrazioni Matematiche Intorno a Due Nuove Scienze.*

Kline, M. (1972). *Mathematical Thought from Ancient to Modern Times*. Oxford University Press.

Peano, G. (1889). *Arithmetices Principia, Nova Methodo Exposita.*

Velleman, D. (2019). *How to Prove It* (3rd ed.). Cambridge University Press.

Wallis, J. (1656). *Arithmetica Infinitorum.*

Historical dates and attributions follow standard scholarly consensus as of publication; specialists continue to debate some particulars, especially around pre-literate and non-European sources.

Appendix Q: A Reader's Discussion Guide

For study groups, classrooms, or a reading partner.

These questions go beyond each chapter's Reflection Questions — they're meant to be talked through, not just written out.

Part One: Foundations (Chapters 1–4)

61. Chapter 1 argues that rote learning and understanding produce different kinds of knowledge. Have the group each describe one thing they know by rote and one thing they understand, and discuss what actually made the difference.
62. Chapter 2 claims the leap from "three sheep" to "the number 3" is one of the most important ideas in mathematical history. Ask the group: what's the modern equivalent of that leap — a moment where a subject stopped being about specific objects and became about abstract structure?
63. Chapter 3's proof that N has no largest element feels almost too simple to be worth proving. Debate: is a proof of an "obvious" fact ever a waste of time? What did the group actually get out of seeing it proven formally?
64. Chapter 4 traces real historical resistance to negative numbers among serious mathematicians. What's a mathematical or

scientific idea today that still meets similar resistance, and why?

Part Two: Structure and Proof (Chapters 5–7)

65. Chapter 5 shows that commutativity and associativity fail for subtraction, division, and matrix multiplication. Have each member name one more everyday "rule" that seems universal but actually only applies in specific contexts.
66. Chapter 6 distinguishes a real proof from a "convincing explanation." Pick a claim from daily life (a news headline, an advertisement) and analyze together whether the argument for it is closer to a proof or to a convincing story.
67. Chapter 7 proves the sign rules instead of just asserting them. Ask the group to each explain, without notes, why negative times negative is positive — then compare explanations. Where did they differ?

Part Three: Application and Synthesis (Chapters 8–10)

68. Chapter 8's Diophantine equations require whole-number answers only. Brainstorm as a group: what other real-world problems secretly require integer, not just numerical, solutions?
69. Chapter 9 argues generalization is a "superpower." Ask each member to generalize a rule of thumb from their own work or hobbies, the way Gauss generalized his pairing trick.

70. Chapter 10's locker puzzle hides a deep result inside a playful setup. What does that suggest about where else in life a "puzzle" might be a real theorem in disguise?

A Closing Question for the Whole Book

Reread the closing lines of the Conclusion. As a group, discuss: has your definition of "understanding" a mathematical idea changed over the course of this book? If so, how would you describe the difference now, compared to before you started?

Appendix R: Index

Page numbers refer to the section where each term is introduced or most fully discussed.

www.ingramcontent.com/pod-product-compliance
Lightning Source LLC
LaVergne TN
LVHW010732061126
851857LV00014B/1441